Published by
YOUNG READERS PRESS, INC.
New York

By JENE BARR
and CYNTHIA CHAPIN

Pictures: P. J. Hoff

What Will the Weather Be?

Picture Dictionary

barometer

hail

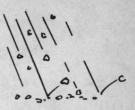

clouds

rain gauge

forecast

THE WEATHER

Fair, low tonight in upper 30s. Tomorrow mostly sunny, high in 60s. West to southwest winds 10-15 m.p.h. Wednesday fair and mild.

rocket

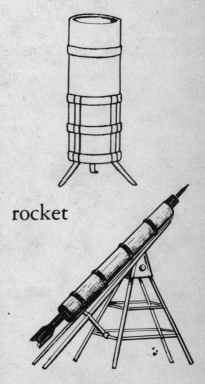

© 1965 by Albert Whitman & Company. This Young Readers Press, Inc., edition is published by arrangement with Albert Whitman & Company.

1st printing .... September, 1970
Printed in the U.S.A.

satellite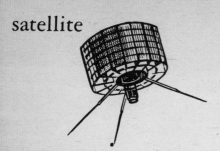

weather map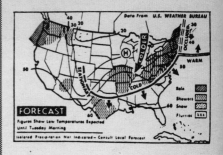

thermometer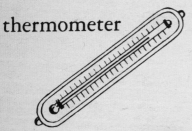

weather vane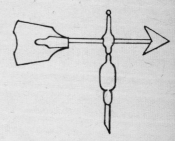

tornado

wind gauge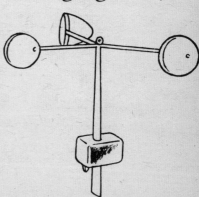

weather balloon

Will it rain or snow?
Or will it be warm and sunny?
The TV man stands at a big map.
He shows where it is raining,
how the wind is blowing,
and where it is warm
and where it is cold.

How does the TV man know
so much about the weather?
He gets weather reports
from many helpers.
These helpers are weathermen.
They work at many stations.

The United States Weather Bureau
has many weather stations.
There are weather stations at airports.
Some ships at sea are weather stations.
There are weather stations
on some mountain tops, too.

Here is Weatherman Bill in the weather station at the airport.
He is a weather forecaster.
He watches the weather and uses weather instruments.
He has a thermometer to tell how hot or cold it is.
This is temperature.

He has a barometer to show
how much the air weighs.
Our earth has air all around it.
What happens in the air
makes the weather.

Rain, snow, wind, and hail
are some kinds of weather.
The weatherman has an instrument
to measure the moisture in the air.
This is humidity.

He has a weather vane to show
which way the wind blows.
The north wind blows from the north.
The south wind blows from the south.
Weatherman Bill says,
"Wind is air going someplace
in a hurry."

How fast is the wind blowing?
On top of a pole, three little cups
spin around.
This instrument tells how fast
the wind blows near the ground.

How about the air high above the earth?
Weatherman Bill sends up
a weather balloon.
This balloon is about as tall as a man.
It carries instruments that send back
signals to the weather station.

What's another way of learning
about the weather? Rockets!
Rockets can fly faster and higher
than balloons.
Weathermen send up rockets
in wide-open places—
over oceans and seas,
and above desert land.

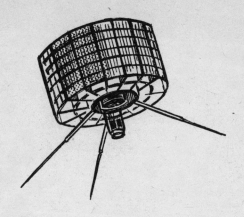

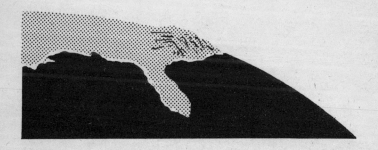

Up high there are weather satellites
that orbit the earth.
These satellites carry instruments
that tell about the weather.
Some have TV cameras that send down
pictures of clouds to weather stations.
Weathermen also use radar to find storms.

And there are "Hurricane Hunters"
who fly into storms!
A hurricane is a big wind.
Often it comes with heavy rain.
How fast is the wind moving?
About 100 miles an hour!

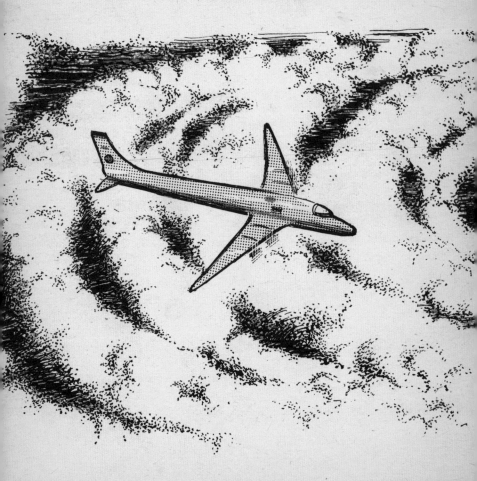

How the plane shakes and shudders!
The hurricane hunters fly
into the "eye" of the hurricane!
Over the radio, these men send news
about the hurricane to weather stations.

RADAR REPORT

Weathermen have other instruments,
and you have them, too. Eyes!
Weatherman Bill looks up at the clouds.
Most clouds are made up
of tiny drops of water.
Some clouds are high up in the sky.
These may be made up of tiny bits of ice.
How light and fluffy these clouds are!

Some clouds are dark and hang low.
These dark clouds may come
with thunderstorms.
After the rain, Weather Man Bill
looks at the rain gauge
to see how much rain fell.

Weatherman Bill gathers all
he has found out about the weather—
the temperature, the humidity,
the rain and snow and wind,
and how the weather fronts are moving.
It's a warm front when warm air pushes
the cool air away.
It's a cold front when cold air
pushes the warm air away.

Weatherman Bill makes a weather report.
Weathermen from other stations make reports.
They send their reports to stations near and far away and to a large office near Washington, D.C.

In this large office the men draw
weather maps.
The maps have weather signs
and numbers on them.
These maps are sent over
electric machines to weather stations
all over the United States.

The weather reports and maps help the forecaster make weather forecasts.

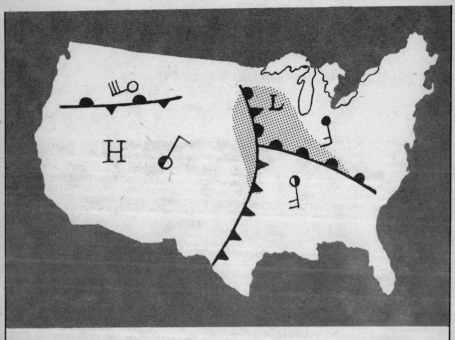

▲▲▲▲ COLD FRONT ●▲●▲ OCCLUDED FRONT

●●●● WARM FRONT L LOW PRESSURE CENTER

▲●▲● STATIONARY FRONT H HIGH PRESSURE CENTER

 AREA OF RAINFALL

◦ CLEAR WITH WIND FROM SOUTHWEST AT 30 MILES PER HOUR

● CLOUDY WITH WIND FROM SOUTHEAST AT 15 MILES PER HOUR

◐ PARTLY CLOUDY WITH WIND FROM SOUTHEAST AT 20 MILES PER HOUR

◐ PARTLY CLOUDY WITH WIND FROM NORTHEAST AT 10 MILES PER HOUR

What will the weather be?
Farmers and fishermen, storekeepers,
people on ships at sea—
everyone wants to know
what the weather will be
tomorrow and even next week.

A farmer listens to the weather forecast over the radio.
He wants to know when he can plant or gather the crops.

At the airport, a pilot studies
the weather reports and forecasts.
He wants to keep away from bad storms.
A truck driver calls the Weather Bureau.
He wants to know if it will snow.

Father and Mother read
the weather report in the newspaper.
Will it be a nice day for the picnic?
Bob and Betsy take their swim suits
so they can go swimming.
Father takes his fishing pole
so he can go fishing.
Mother takes her sun umbrella
so she can just sit and rest.

Day and night, people want news
about the weather.
Forest Rangers listen to
fire-weather forecasts.
Fires start easily when forests are dry.

A storm warning!
A tornado warning!
A flood warning!
High waves!
People must have time to get ready
for the storm.

Some day we may know for sure
what the weather will be.
Scientists are learning how to
"make weather."
At certain times and in some places
men can make more rain fall.
Sometimes men can make snow, too.

Do you want to go skiing or swimming?
Some day you may be able to choose
the weather you want.
Then the weatherman will be
"a weather planner."

But where is the TV man?
He has looked at the weather forecasts.
Now he stands at a big map.
He talks about the weather
all over the United States.
Best of all, the TV man tells
what the weather will be
right where you live!